Beautiful Birds

EAGLES

By Bray Jacobson

Gareth Stevens
PUBLISHING

Please visit our website, www.garethstevens.com. For a free color catalog of all our high-quality books, call toll free 1-800-542-2595 or fax 1-877-542-2596.

Library of Congress Cataloging-in-Publication Data

Names: Jacobson, Bray, author.
Title: Eagles / Bray Jacobson.
Description: Buffalo, New York : Gareth Stevens Publishing, [2024] | Series: Beautiful birds | Includes index. | Audience: Grades K-1
Identifiers: LCCN 2023023602 (print) | LCCN 2023023603 (ebook) | ISBN 9781538292471 (library binding) | ISBN 9781538292464 (paperback) | ISBN 9781538292488 (ebook)
Subjects: LCSH: Eagles–Juvenile literature.
Classification: LCC QL696.F32 .J33 2024 (print) | LCC QL696.F32 (ebook) | DDC 598.9/42–dc23/eng/20230525
LC record available at https://lccn.loc.gov/2023023602
LC ebook record available at https://lccn.loc.gov/2023023603

First Edition

Published in 2024 by
Gareth Stevens Publishing
2544 Clinton St.
Buffalo, NY 14224

Editor: Kristen Nelson
Designer: Andrea Davison-Bartolotta

Photo credits: Cover rbrown10/Shutterstock.com; p. 1 Al Carrera/Shutterstock.com; p. 5 Locomotive74/Shutterstock.com; p. 7 Skyler Ewing/Shutterstock.com; pp. 9, 24 (talon) Alexandre Boudet/Shutterstock.com; p. 11 (bald eagle) Christina Dutkowski/Shutterstock.com; p. 11 (golden eagle) J Edwards Photography/Shutterstock.com; p. 13 jgorzynik/Shutterstock.com; p. 15 Jed Packer/Shutterstock.com; pp. 17, 19, 24 (cliff, nest) Danita Delimont/Shutterstock.com; p. 21 Ronnie Howard/Shutterstock.com; p. 23 wendy blanchard/Shutterstock.com.

Printed in the United States of America

CPSIA compliance information: Batch #CW24GS: For further information contact Gareth Stevens, New York, New York at 1-800-542-2595.

Contents

I see a bird!
It is an eagle.

It has feathers.
It has large wings.

It has claws.
These are talons.

There are more than 70 kinds. Only two kinds live in North America.

They hunt for food.
Some catch fish.

They eat dead animals!

Nests are in tall trees.
They can be on cliffs!

Nests are made
of sticks.
Parents use the same
nest for years!

Mothers lay two eggs at a time.

Both parents feed
the babies.

Words to Know

cliff

nest

talon

Index

Beautiful Birds

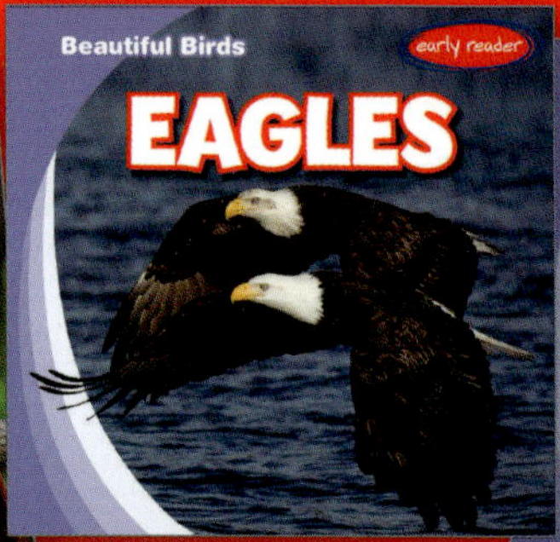

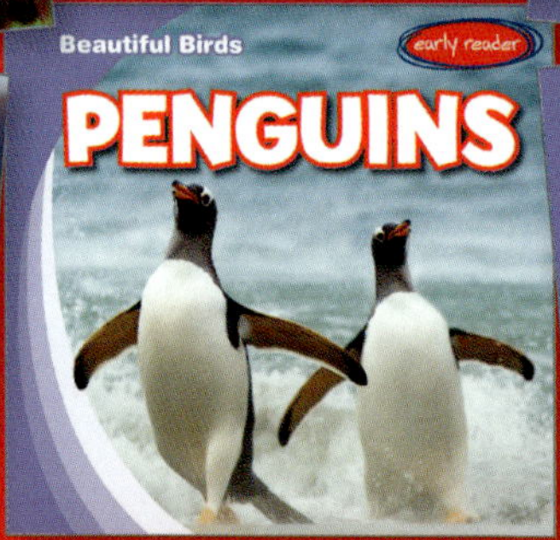

ISBN: 9781538292464

9 781538 292464

Gareth Stevens
PUBLISHING